João Eduardo Campos Silva
Vagner Marcelo Gomes

Biopolymer obtained from starch

João Eduardo Campos Silva
Vagner Marcelo Gomes

Biopolymer obtained from starch

Polymer chemistry

ScienciaScripts

Imprint

Any brand names and product names mentioned in this book are subject to trademark, brand or patent protection and are trademarks or registered trademarks of their respective holders. The use of brand names, product names, common names, trade names, product descriptions etc. even without a particular marking in this work is in no way to be construed to mean that such names may be regarded as unrestricted in respect of trademark and brand protection legislation and could thus be used by anyone.

Cover image: www.ingimage.com

This book is a translation from the original published under ISBN 978-613-9-62298-6.

Publisher:
Sciencia Scripts
is a trademark of
Dodo Books Indian Ocean Ltd. and OmniScriptum S.R.L publishing group

120 High Road, East Finchley, London, N2 9ED, United Kingdom
Str. Armeneasca 28/1, office 1, Chisinau MD-2012, Republic of Moldova, Europe
Printed at: see last page
ISBN: 978-620-7-72068-2

SUMMARY

DEDICATORY

I dedicate this work to my family, especially my parents Joâo and Evanir, my sister Keylla, my brother-in-law Bruno and my nephew Pedro.

ACKNOWLEDGMENTS

I would first like to thank God for having been present at every moment of my life and always lighting my way.

To my parents for always motivating me in my studies, never giving up despite times of difficulty.

To my advisor Carlos Eduardo B. Cassini for his support and ideas during the development of the TCC.

I would like to thank the teachers who directly or indirectly answered questions that were important to me during the course.

I would like to thank the great friends I made during my degree, Cristiane, Luciana and Vagner, who were always there for me in times of difficulty.

To my friends André, Gabriel, Joâo Victor, Leandro and Livia for their friendship and for always being present in my life.

PRESENTATION

Polymers are responsible for a large part of environmental degradation, as they are derived from petroleum, i.e. they are non-renewable sources and take hundreds of years to decompose. The general objective of this work is to produce a biopolymer from cassava starch. The specific objectives are to extract the starch from cassava, promote the acid hydrolysis of the starch using 0.1M hydrochloric acid, develop the biopolymer and analyze the decomposition time of this biopolymer. To extract this starch, a number of unit operations were carried out during the extraction process, the main ones being crushing, straining, decanting and drying the starch, which was then weighed into 5 grams and heated together with water, During this heating, 6 ml of 0.1M hydrochloric acid was added in order to promote acid hydrolysis (breakdown) of the starch, and in order for this biopolymer to take on a more malleable form, a plasticizing agent had to be added, to which 4 ml of glycerol (glycerin) was added. After producing this biopolymer, it was observed that it had a smooth texture and good elasticity. After observing these characteristics visually, the behavior of this biopolymer was observed in aquatic, dry terrestrial and wet terrestrial environments, After this analysis it can be concluded that from the second week this biopolymer began to decompose and suffered its total decomposition in just 8 weeks in places where water was present, with these results it was concluded that this biopolymer decomposed quickly satisfying the hypothesis of the work which is the development of a biopolymer of rapid decomposition.

Key words: biopolymer, starch, decomposition, hydrolysis, cassava.

1. INTRODUCTION

Plastics produced synthetically using macromolecules derived from petroleum take a long time to degrade in the environment, leading to exacerbated environmental pollution. The first plastics appeared in the 19th century with the vulcanization process developed by Charles Goodyear when he improved natural rubber, making it more resistant. Other important plastics are Polyethylene Terephthalate (PET), Polyvinyl Chloride (PVC), Polystyrene (PS), Polypropylene (PP), High Density Polyethylene (HDPE), Low Density Polyethylene (LDPE), Nylon, Polyester, among others.

Nowadays, plastic is present in almost every type of market, such as:

- Food

- Aeronautics

- Automotive

- Cosmetics

- Household

- Pharmacists

- Industrial

- Pipes

Due to the high demand from the market and the fact that plastic is produced from a non-renewable source such as oil, it generates a huge amount of waste that will degrade for a long time, making it difficult to dispose of. The production of plastics requires a large industrial process to transform oil into polymers and thus give rise to plastics. Environmental pollution in large cities is due to the accumulation of plastic bags and PET bottles, causing floods and leaving the city looking dirty.

With the emergence of green chemistry, also known as sustainable chemistry, some companies are developing products that seek to reduce environmental impact. Researchers have now succeeded in developing green polyethylene made from sugar cane, which has the same characteristics as petroleum-based polyethylene. New sources for the production of plastics are emerging, such as starch, which is already being used to produce plastics from corn starch, and companies are already developing disposable cups from cassava starch.

Starch can be found in a variety of sources such as corn, potatoes, sweet potatoes and cassava, but in order to develop plastics from starch, it must be broken down by enzymatic hydrolysis, in which the enzymes are responsible for breaking down the starch, or acid hydrolysis, in which the acid is responsible for breaking down the starch molecule and thus, through industrial processes, with the help of plasticizing agents, this starch is transformed into plastic. This biopolymer breaks down quickly, in about 4 to 8 weeks.

- Justification

Searching for a way to produce a biopolymer in cassava that decomposes quickly, unlike the plastic produced from petroleum, which takes years to decompose.

- .2 Hypothesis

The biopolymer has a decomposition period of a few weeks.

2. OBJECTIVE

2.1. General objective

To develop a biopolymer from starch that breaks down quickly in the environments tested.

2.2. Specific objective

- Extracting starch from cassava

- Promote acid hydrolysis of starch

- Developing the biopolymer from cassava

- Observe the decomposition behavior of this biopolymer.

3. THEORETICAL BASIS

3.1. History of plastics

According to the Aurélio dictionary, plastic is a synthetic material that is highly malleable and easily transformed by the use of heat and pressure.

Plastics date back to the 19th century when in 1839 the American Charles Goodyear developed a process called vulcanization.

This process consists of adding predetermined amounts of sulphur, this amount varying according to its subsequent application. The addition of sulfur makes the rubber more resistant to heat, thus making it of great importance in the automotive industry for use in car tires.

Figure 1 below shows the vulcanization process:

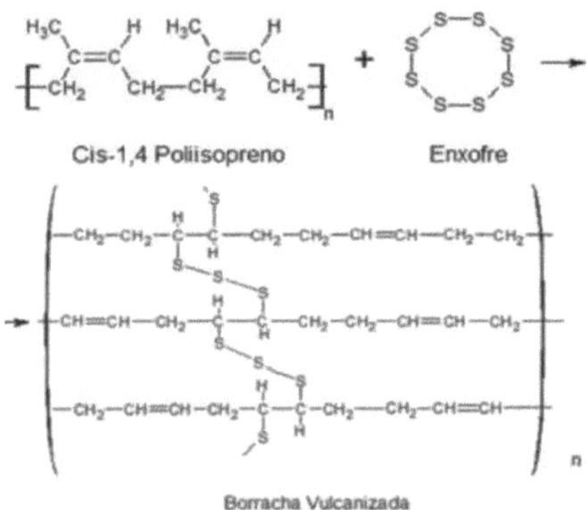

Figure 1-Vulcanization of rubber.

Source: http://www.mundoeducacao.com/quimica/vulcanizacao-rubber.htm

In 1909 Leo Baekand discovered Bakelite, which was a synthetic resin. Due to the characteristics of this polymer, it has many applications such as:

- Pot handles
- Socket switches

- Billiard balls

Figure 2 below shows how Bakelite is formed:

Figure 2-Polymerization of PF

Source: MORASSI, 2011, p.63

After the emergence of Bakelite in the 20th century, new research was carried out and new plastics were developed:

- Polyethylene (PE)

- Polystyrene (PS)

- Polyethylene Terephthalate (PET)

- Polyvinyl chloride (PVC)

Nylon

Polypropylene (PP)

These plastics come from oil and are therefore non-renewable, taking hundreds of years to decompose in the environment, thus causing environmental pollution.

Based on the idea that conventional plastics take a long time to break down, researchers have been carrying out new research using green chemistry.

3.2. Types of plastics

Plastics are formed by the polymerization reaction in which monomers are joined together to form macromolecules called polymers.

Polymers can be divided into thermoplastics, thermosets or thermosets, elastomers and biopolymers. These plastics differ in the way they are produced:

3.2.1. Thermoplastics

When treated with temperature and pressure, they soften and take on the shape of the mold in which they are placed. When they are heated under pressure again, the process begins again, making them recyclable. These plastics are soft and ductile and include: PE, PP, PVC and PET (BARROS, 2011).

3.2.1.1 Polyethylene (PE)

3.2.1.1. Low density polyethylene (LDPE) and linear low density polyethylene (LLDPE)

They are materials with low electrical and thermal conductivity and are resistant to the action of chemical and non-toxic products.

It is a flexible, light and transparent plastic.

They are used in food packaging, personal hygiene products and supermarket bags.

3.2.1.1.2. High-density polyethylene (HDPE)

They are opaque materials due to their high density; they are light, waterproof, rigid and have good mechanical resistance.

Due to its characteristics, it is used in packaging for cleaning products.

Figure 3 below shows the polymerization reaction of ethylene to form polyethylene.

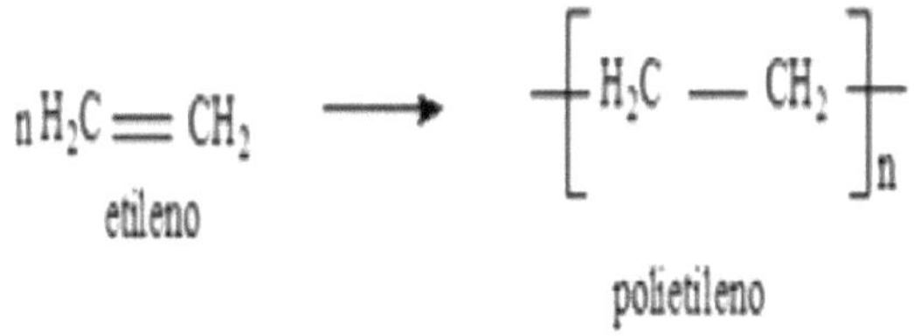

$$n\,H_2C = CH_2 \longrightarrow \left[H_2C - CH_2 \right]_n$$

etileno

polietileno

Figure 3-Polyethylene polymerization reaction (Source: MORASSI, 2011, p.13)

3.2.1.2 Polypropylene (PP)

This polymer has low impact resistance, so it can be improved with a copolymerization reaction. Polypropylene has low permeability to gases and little permeability to water vapors.

1.1.1.1.1. Homopolymer polypropylene

It is a material resistant to high temperatures and can therefore be sterilized. It has good chemical resistance and few organic solvents are capable of solubilizing it. It is used in auto parts, food packaging and fibers (MORASSI, 2013).

1.1.1.1.2. Copolymer polypropylene

It is a transparent material that is more flexible and resistant than homopolymer. When modified with elastomers, it is more resistant to impact and has high mechanical strength, making it a raw material for household utensils such as jars and packaging (MORASSI, 2013).

Figure 4 below shows the polymerization reaction of propylene to form polypropylene.

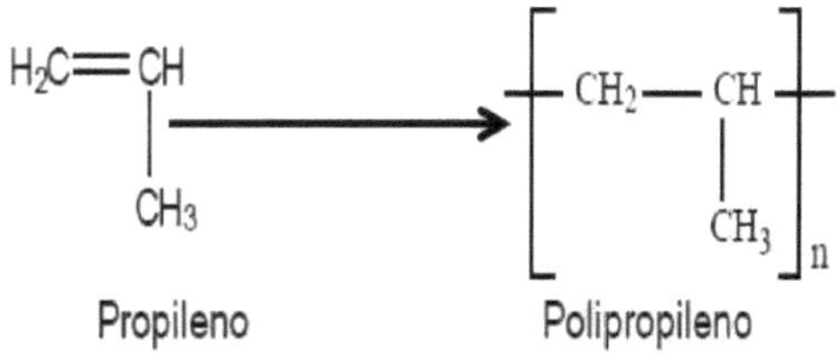

$$H_2C = CH \atop | \atop CH_3 \longrightarrow \left[CH_2 - CH \atop | \atop CH_3 \right]_n$$

Propileno   Polipropileno

Figure 4-Polymerization reaction of polypropylene

Source: MORASSI, 2011, p.16

1.1.1.3. Polystyrene (PS)

It is a rigid, hard and transparent polymer with low cost, good moldability, low moisture absorption, good electrical insulation and good chemical resistance (MORASSI, 2013).

Polystyrene is divided into three groups:

1.1.1.3.1. Crystal polystyrene

Amorphous, hard, shiny material with a high refractive index. Used in low-cost articles (PASSATORE, 2013).

1.1.1.3.2. Expanded polystyrene

A blowing agent is applied during its production, making the material foamy and giving it excellent acoustic and thermal properties. They are used in food packaging, thermal insulation and construction. (MORASSI, 2013)

1.1.1.3.3. High-impact polystyrene

Its formulation contains 5 to 10% elastomer, added by mechanical mixing or directly in the polymerization process. This elastomer makes it widely used in the manufacture of household utensils and toys (PASSATORE, 2013).

Figure 5 below shows the polymerization reaction of styrene to form polystyrene.

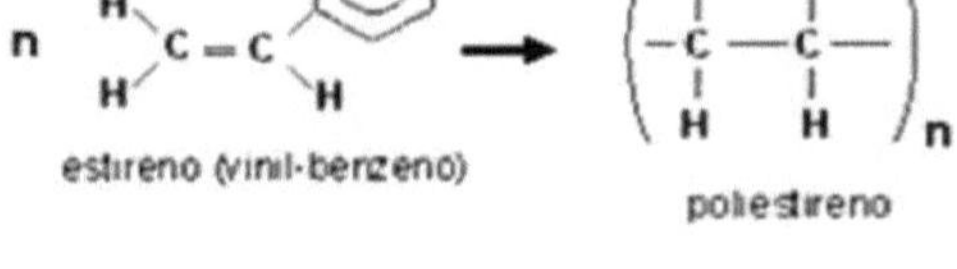

Figure 5-Polystyrene polymerization reaction

1.1.1.4. Polyvinyl chloride (PVC)

It is not a 100% petroleum-based material, as it contains 57% chlorine and 43% petroleum. PVC is used to make boxes and roof tiles. With the addition of plasticizing agents, it becomes softer, making it useful in the manufacture of flexible tubes, gloves, shoes and others (PASSATORE, 2011).

PVC is divided into two groups:

1.1.1.4.1. Flexible PVC

Widely used in electrical insulation, hoses, films for food packaging.

1.1.1.4.2. Rigid PVC

It is widely used in the chemical industry due to its high resistance to corrosive products and in civil construction in pipes and coatings (MORASSI, 2013).

Figure 6 below shows the polymerization reaction of vinyl chloride to form polyvinyl chloride or PVC.

Figure 6- Polymerization of PVC Source: MORASSI, 2011, p.26

1.1.1.5. Polyethylene Terephthalate (PET)

It has a high melting temperature and high hydrolytic stability due to the presence of aromatic rings in the main chain. PET applications include textile fibers, packaging, bioriented films and others (Româo et al, 2009).

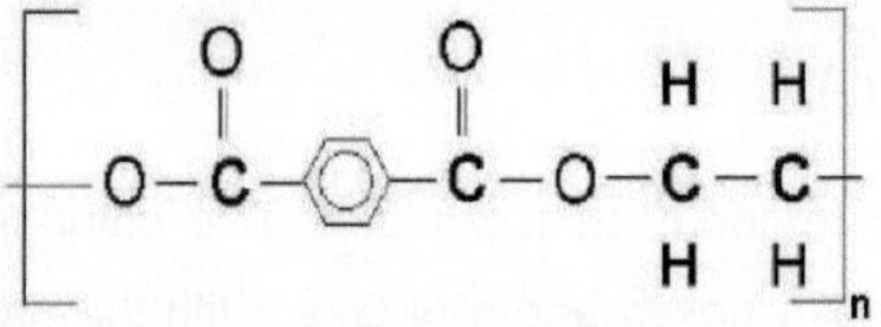

Figure 7-Polymerization of PET

Source: PASSATORE, 2013, p.62

3.2.2. Fixed term

When heated, they take the shape of the mold in which they were placed, but when heated again under high pressure there is no effect, making them insoluble and non-recyclable. The main polymers in this group are: Polyurethane (PU) and Polyphenol (PF) or Bakelite.

3.2.2.1. Polyurethane (PU)

It is obtained from paraphenylene diisocyanate and ethylene glycol (1,2-ethanediol). It is characterized by its resistance to abrasion and heat, and is used in insulation, internal lining of clothing, among others. When it undergoes the hot expansion process, a foam is formed whose hardness can be controlled according to its application (PASSATORE, 2011).

Figure 8 below shows the polymerization reaction to form polyurethane or PU.

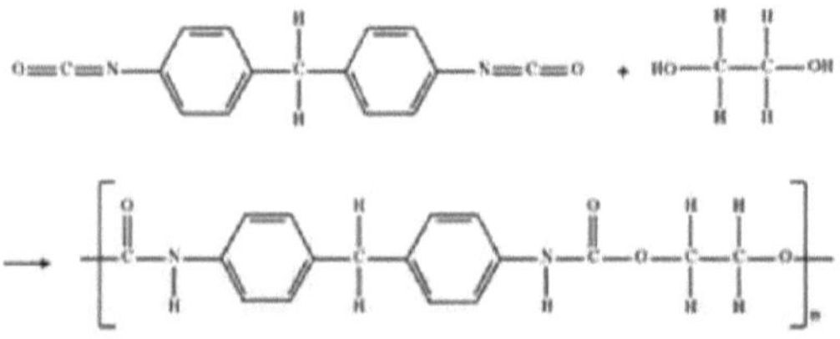

Figure 8-Polymerization of PU

Source: MORASSI, 2011, p.58

3.2.2.2. Polyphenol - Bakelite (PF)

Phenol-formaldehyde is a resin produced by the reaction of phenols with formaldehyde; phenolic resins are used for the production of printed circuit boards, foundry molding, sandpaper, among others, and are the oldest synthetic polymers. (MORASSI, 2011)

Figure 9 below shows the polymerization reaction for to form polyphenol.

Figure 9-Polymerization of PF

Source: MORASSI, 2011, p.63

3.2.3. Elastomers

They are also known as rubbers and once melted they cannot be melted again.

Natural rubber (latex) and synthetic rubber (SBR) belong to the elastomer class.

3.2.3.1. Natural rubber or latex

It is a natural polymer obtained from the sap of the rubber tree. Due to its elasticity, it is used in shock absorbers, car cushions and surgical gloves (MORASSI, 2013).

Figure 10 below shows the polymerization reaction for the formation of natural rubber or latex.

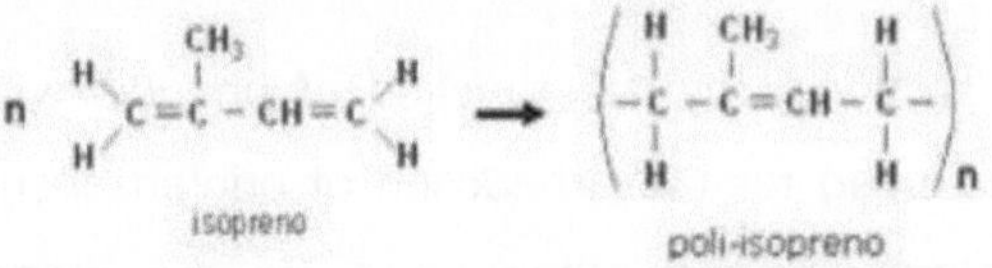

$$n \quad \underset{H}{\overset{H}{C}}{=}\underset{\overset{|}{H}}{\overset{\overset{CH_3}{|}}{C}} - CH = C \underset{H}{\overset{H}{\big\langle}} \longrightarrow \left(-\underset{\overset{|}{H}}{C} - \underset{}{\overset{\overset{CH_2}{|}}{C}} = CH - \underset{\overset{|}{H}}{C} - \right)_n$$

isopreno   poli-isopreno

Figure 10-Polymerization of natural rubber

Source: MORASSI, 2011, p.104

3.2.3.2. Synthetic rubber

There are several types of synthetic rubber:

1.1.3.2.1. NBR: Nitrile Rubber

Obtained from the copolymerization of butadiene with acrylonitrile. It is used to produce parts that will come into contact with oil or oil derivatives. It has low volume variation and good tensile strength when exposed to petroleum products.

1.1.3.2.2. SBR: Styrene butadiene rubber

Obtained from the copolymerization of styrene with butadiene. It is used in products that do not require mechanical resistance and is characterized by its low cost.

1.1.3.2.3. IIR: Isobutylene isoprene rubber

Rubber with low gas permeability. Used in tire tubes (MORASSI, 2013).

3.3. Plastic's environmental problem

Plastic is a synthetic polymer and its waste generation comes directly from anthropogenic actions (NUCCI, 2010 apud CHESHIRE et al., 2009).

This generation of waste and its incorrect disposal causes damage to the terrestrial and aquatic environment.

In the aquatic environment, their effects are very difficult to observe, but they are certainly lethal and can lead to suffocation, obstruction of the digestive tract, as well as entanglement in the animal, making it difficult for it to move (NUCCI, 2010 apud BARBIERE, 2009). Among the animals most affected are the seabirds that ingest these plastics, as they are mistaken for prey, so the animals that don't die transmit the toxins to their nests. (NUCCI, 2010)

With these problems in mind, ecologists have come up with some very convincing new ideas to solve the problem of waste. We will have to adopt new attitudes that involve: reducing consumption, reusing materials and recycling. This new attitude is an ever-increasing demand from modern societies that aspire to rational growth, based on so-called sustainable development. (PIATTI & RODRIGUES, 2005).

3.4. Recycling

Plastics belonging to the thermoplastic group can currently be recycled. This recycling can take place in three ways:

3.4.1. Mechanics

This is the most common method used in Brazil. It consists of converting plastics from industries or discarded by households into small granules, which are then used to produce new materials such as plastic bags, flooring, hoses and so on.

Figure 11 below shows more clearly how the mechanical recycling process is carried out:

Figure 11-Mechanical recycling.

Source:

3.4.2. Chemistry

This is a more elaborate process, as it consists of reprocessing plastics and converting them into basic petrochemical materials, which serve as raw materials for high-quality materials. This process is more tolerant of impurities, but it is more expensive and requires large quantities of plastics.

Figure 12 below shows a flowchart of chemical recycling:

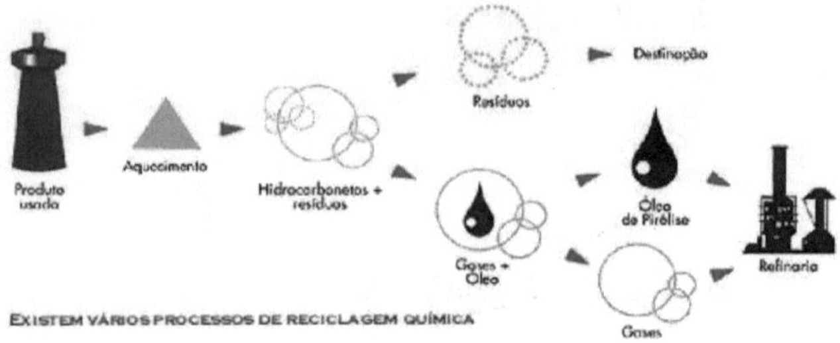

Figure 12-Chemical recycling

Source:

http://www.plastivida.org.br/2009/images/Etapas_ReciclaQuimica.gif

3.4.3. Energy

A process that converts plastic into thermal and electrical energy. This process consists of the incineration of plastics and the energy is obtained through the calorific power it possesses. Energy recycling is a very important process, as it creates energy matrices that greatly benefit cities that don't have a suitable place to dispose of urban waste.

Figure 13 below shows the energy recycling flowchart:

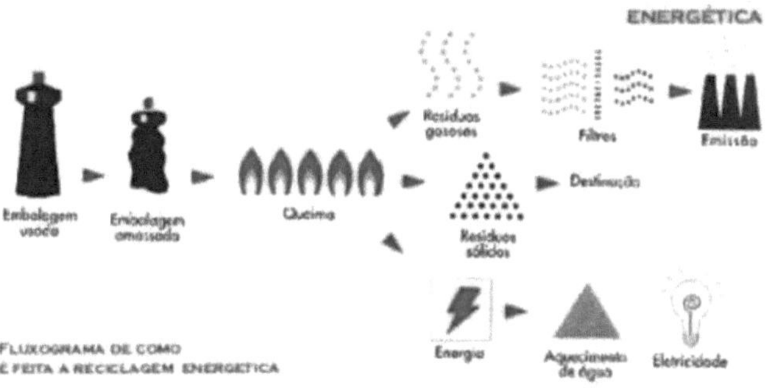

Figure 13-Energy recycling

Source: https://oplastico.files.wordpress.com/2013/10/reciclagem-do-plastico2.jpg

3.5. Biopolymers

They are polymers or copolymers produced from raw materials from renewable sources, such as corn, sugar cane, cellulose, potatoes, cassava and others. Renewable sources are sources that have a shorter life cycle than non-renewable sources such as oil. The fact that leads to great interest in biopolymers is that they do not cause as much environmental damage as oil extraction and refining. In addition to the problems arising from the extraction and refining of oil, there is the fact that these plastics are not biodegradable, i.e. they take hundreds of years to break down, generating an exacerbated amount of plastic waste in the environment.

With the advance of research and the growth of technology, there are biopolymers that have great potential for replacing petroleum-based polymers. However, despite these possible replacements, biopolymers have some limitations, such as their processability and their use as a finished product. Researchers have been developing ways to increase their resistance, mechanical properties and degradation rates.

This is how biodegradable polymers come about, where they are degraded by the action of microorganisms such as fungi and bacteria, taking a few weeks to break down under favorable conditions. These are developed from renewable sources such as starch from potatoes, cassava and corn, sugar cane and even cellulose.

The decomposition of biopolymers can occur in three ways:

3.5.1. Biodegradation

This decomposition occurs through the action of microorganism enzymes, which are suitable for breaking the chemical bonds of biopolymers, under suitable conditions of temperature, humidity and pH (BRITO, 2011).

This biodegradation can occur in the presence or absence of oxygen, characterizing aerobic biodegradation (presence of oxygen) or anaerobic biodegradation (absence of oxygen) (BRITO et al, 2011).

3.5.2. Composting

A process in which the decomposition process transforms bioplastics into humus-like compounds, resulting in the production of carbon dioxide, water, minerals and organic matter. In this way, this process gives up carbon dioxide, water, inorganic compounds and biomass to the environment without producing toxic waste (BRITO et al, 2011).

3.5.3. Oxybiodegradation

These plastics consist of polymers that have some additives in their formulation that accelerate their oxidative degradation when in the presence of light or heat. So a plastic that would take around 400 years to decompose takes only 18 months to decompose. These plastics go through two stages of degradation: abiotic, which is accelerated by the catalyst, and biotic, which is accelerated by microorganisms (BRITO et al, 2011).

3.6. Main biopolymers

3.6.1. Green polyethylene

It is a combination of innovation, technology and sustainability.

After years of research, plastics made from ethanol extracted from sugar cane have been developed in Brazil. Unlike oil, ethanol is obtained from sugar cane, which is a renewable source.

Green polyethylene retains the same properties, performance and versatility as conventional polyethylene, making it easier to recycle along with common polyethylenes.

These are already produced as high-density polyethylene (HDPE) and low-density polyethylene (LDPE) and linear low-density polyethylene (LLDPE).

3.6.2. Biodegradable plastic

In the process of biodegradation, organic compounds from the environment are converted into simpler compounds, returning to nature through elemental cycles such as carbon, nitrogen and sulphur. Biodegradable plastic undergoes significant changes in its chemical structure under specific environmental conditions (BONA, 2007).

The development of biodegradable packaging from renewable resources promotes the creation of economic alternatives as well as reducing environmental impacts (BONA, 2007 apud LAROTONDA, 2002).

3.6.3. Poly lactic acid (PLA)

PLA is a thermoplastic polyester made from lactic acid, obtained from renewable sources such as corn, cassava, sugar beet and sugar cane. They are therefore biodegradable, compostable and recyclable plastics, and can also be incinerated.

Products developed from PLA:

- Plastic bags
- Packaging for cosmetics
- Food packaging
- Disposable cups, cutlery and plates.

Advantages:

- Biodegradable

- Compostable

- Can be recycled.

Disadvantages:

- Lack of suitable places for biodegradation

- Most waste goes to landfills or dumps

- Use of edible raw materials for its production

- Large area for planting raw materials

- Fertilizer use in crops.

3.6.4. Thermoplastic starch

Natural starch does not have thermoplastic characteristics due to the strong intermolecular interactions (hydrogen bonds), so in order for this starch to acquire thermoplastic properties, it needs to be heated. And for this starch to melt, a plasticizing agent must be used. There are two processes for destroying the crystalline structure:

Gelatinization: consists of adding solvents that are able to diffuse through the starch granules, dissolving the amylose and amylopectin.

Plasticizer: Occurs at temperatures sufficient to break the hydrogen bonds, causing loss of crystallinity with swelling and breakage of the granule, with the possibility of partial dissolution of the amylose (MIRANDA, 2011).

Products produced from thermoplastic starch:

- Films for food protection

- Children's diapers

- Cotton buds

- Replacing expanded polystyrene

- Toys

- Pens.

Advantages:

- Renewable

- Biodegradable

- Low cost

- High availability

- Processed on conventional plastics machines.

Disadvantages:

- Not very resistant to water

- It needs to be mixed with other plastic elements to improve its resistance, so it is no longer 100% biodegradable.

- It uses edible sources for its production.

3.6.5. Difference between PLA and Thermoplastic starch

Both are obtained from starch, but PLA uses starch as a source for obtaining lactic acid, while starch plastic (thermoplastic starch) uses starch as a raw material, thus maintaining 100% biodegradability.

3.6.5.1 Starch

It is a natural polymer, found in various agricultural sources such as corn, cassava, potatoes, wheat and rice. Starch is a linear molecule in the region where amylose is found and in the form of short chains in which amylopectin is found. (MIRANDA, 2010)

3.6.5.2. Amylose

Amylose is a linear polymer composed of α- 1,4 bonds in the D-glucose units.

The amylose content in starch varies according to the source of origin, with common corn having an amylose concentration of 25 to 28% and cassava having a lower concentration of 17% (SPIER, 2010).

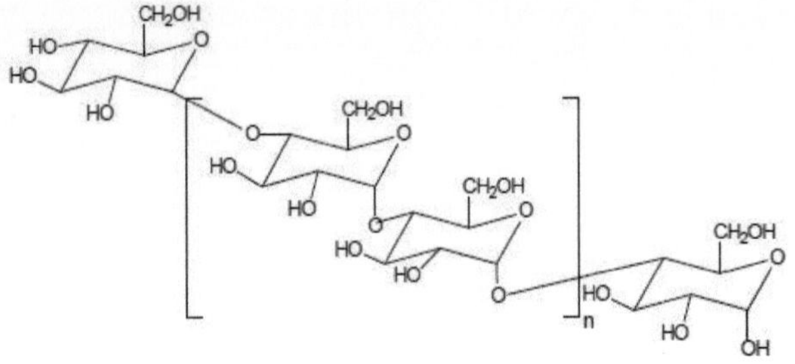

Figure 14-Chemical structure of amylose

(Source: MIRANDA, 2011, p.5)

3.6.5.3.
Amylopectin

Amylopectin has α- 1,4 bonds in the D-glucose units of the main chain, but is characterized by a high degree of branching, with 5 to 6% of α- 1,6 bonds in the D-glucose regions (SPIER, 2010).

Amylopectin has a concentration of 70 to 80% in common corn, potatoes and cassava (MIRANDA, 2011).

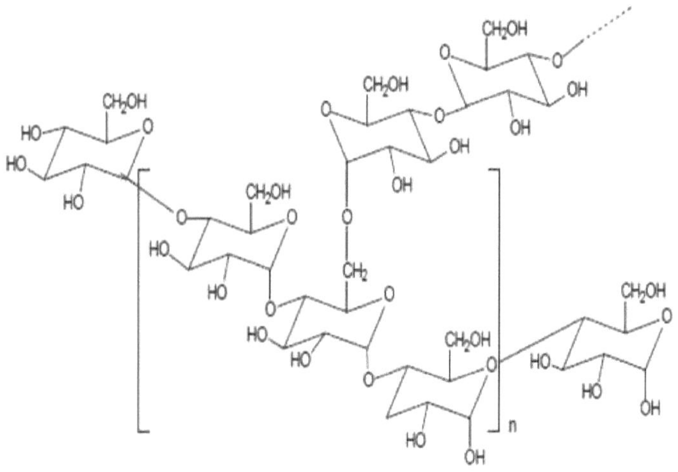

Figure 15-Amylopectin.

(Source: MIRANDA, 2011, p.5)

3.7. Cassava

Scientific name: *Manihot esculenta crantz*

Family: Euphorbiaceae

Known names: Cassava, Macaxeira, and Aipim.

It is an edible root originating in South America and is one of people's main energy foods. (EMBRAPA)

In addition to its high energy content, cassava also contains mineral salts and B vitamins. It should be grown in tropical and subtropical regions with an average temperature of 25°C. It is a drought-tolerant root and has a wide range of climatic and soil adaptations (LAZZARI, CPT - Center for Technical Production).

3.7.1. Manioc Mansa (Common Cassava)

Cassava is used as food for humans and animals.

3.7.2. Wild Cassava

Used for industrial purposes.

When this cassava comes into contact with the hydrochloric acid present in the stomach, it causes acid hydrolysis and releases cyanide acid into the body, which is extremely harmful or even fatal to humans.

3.7.3. Cassava by-products

Cassava by-products are constituent parts of the plant itself. During the processing of cassava to produce starch and flour, the by-products generated can be solid or liquid. Among the main by-products are:

- Leaves:

They have considerable levels of carotenes and vitamin C, but this is lost during the drying process, as is HCN.

- Stem:

It has a 60% starch content.

- Bark:

It is a thin brown layer of cellulose that accounts for 2-5% of the total weight of cassava.

- Pasta:

It is generated during the extraction of starch, as it is soaked in water and has a large volume, so it is around 75% water. The main composition of the residue is around 63% starch, 8% fiber and 2% protein, the rest is divided into small percentages of glucose, phosphorus, calcium, potassium and sterile extract (EMBRAPA, 2010).

3.8. National cassava starch production

Below is a graph showing the variation in national cassava starch production.

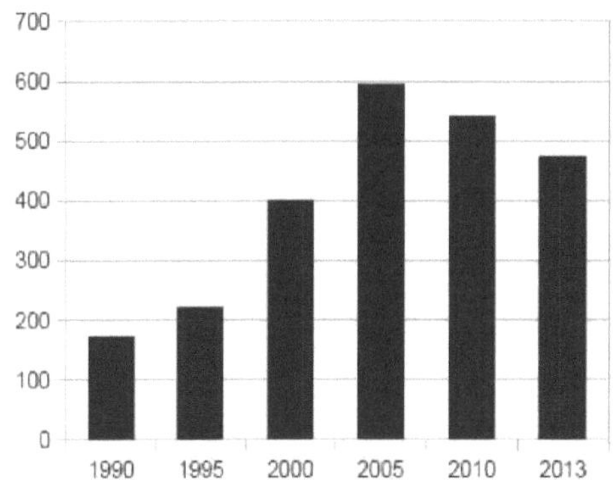

Graph 1- Brazilian cassava starch production in 1000T

(Source: Analysis of the agricultural situation - Cassava 2014/2015 harvest.

SEAB, DERAL, p.9)

Based on the data in the graph above, it can be seen that the period of lowest cassava starch production was in the 1990s, but from 2000 onwards, production began to grow more and reached its peak in 2005. In this period from 2000 to 2013, average production was around 500,000 tons of cassava starch.

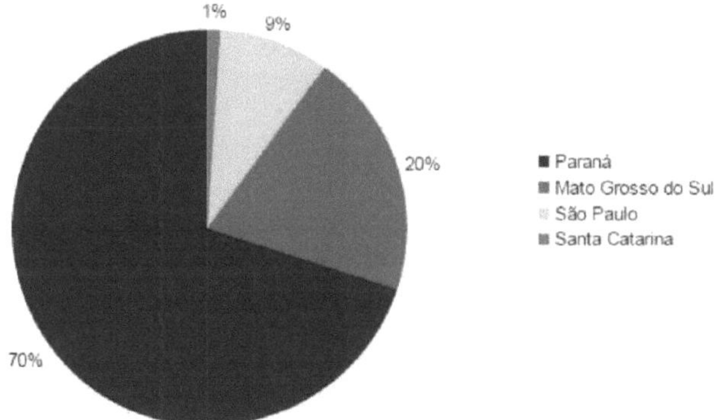

Graph 2- Starch production: Top producers 2013

(Source: Analysis of the agricultural situation - Cassava 2014/2015 harvest.

SEAB, DERAL, p.10)

The chart above shows the largest producers of cassava starch in 2013. This chart shows that the largest producer of starch in Brazil is Paranà; this leadership of the state of Paranà as the largest producer is due to the fact that during this period the northeastern region faced a great period of drought, thus making Paranà the largest producer.

The table below shows the largest exporters of cassava starch from 2010 to 2013.

Table I - Main Exporting States - 2010/2013

STATES	2010		2011		2012		2013	
	(t)	U$$	(t)	USS	(t)	USS	(t)	UtS
PARANÂ	2201	2.273	2.501	2.192	3.596	3.335	2.642	2.591
SAO PAULO	1 528	1.034	1.100	691	409	547	147	230
MATO GROSSO DO SUL	1.323	1.164	1.455	1.334	1.431	909	1.301	892
SANTA CATARINA	741	689	463	600	1.712	1.392	667	682
OTHER	191	242	1.243	754	115	166	40	58
BRAZIL	5 984	5.402	6.762	5.571	7.262	6.309	4.797	4.453

(Source: Analysis of the agricultural situation - Cassava 2014/2015 harvest - SEAB, DERAL, p. 13)
SEAB, DERAL, p. 13)

As shown in Graph 2, in which Paranâ was the largest producer of cassava starch in 2013, Paranâ was also the largest exporter of cassava starch during the analysis period between 2010 and 2013, with an average of 2735 tons exported during the period.

3.9. Products made from cassava

The cultivation of cassava provides various starch products, this starch is called starch and is used for food and industrial purposes, such as tapioca, starch, cookies and other food products. Cassava starch is also used as a raw material for the development of paints, medicines and other products. Currently, with the search for renewable sources, starch is being used to develop plastic bags, disposable cups and household utensils.

1.10. Cassava economy

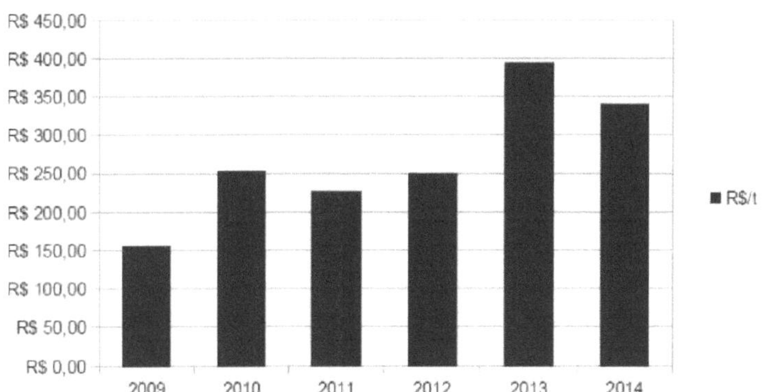

Graph 3 - Evolution of cassava root prices received by producers in Paraná.

Source: Analysis of the agricultural situation - Cassava 2014/2015 harvest.

SEAB, DERAL, p. 17

The graph above shows the evolution of cassava root prices from 2009 to 2014. The highest prices were in 2013 and 2014, with an average price of 270.00R$ per ton of cassava root.

Table 2 - Cassava imports in Brazil

	2012		2013	
	US$	Tons	US$	Tons
United States	39.999	18	105.136	75
Paraguay	6.601.259	27.363	11.154.606	26.668
Vietnam	-	-	104.192	18

| Thailand | - | - | 9.514 | 238 |
| Total | 6.641.258 | 27.382 | 11.373.448 | 26.999 |

(Source: Panorama of the cassava market - SEBRAE, 2014, p.4)

Although Brazil is a major cassava producer, it had to import cassava from other countries in 2012 and 2013.

1.11. Consumption

The consumption of cassava is characterized by the products derived from it, such as flour, sour powder and starch, most of which are used for human consumption.

Nowadays, cassava is also being consumed in industrialized forms such as frozen cassava sticks and cassava chips.

Other branches of industry are using cassava to produce biodegradable plastics.

Table 3 - Per capita household food consumption of cassava

	CASSAVA	CASSAVA flour	CASSAVA STARCH
BRAZIL	1,77	5,33	0,77
NORTH	2,78	23,54	1,56
NORTH EAST	1,35	9,67	1,44
SOUTHEAST	0,99	1,17	0,36
SOUTH	4,12	0,81	0,30
CENTRO-OESTE	2,03	1,29	0,65

(Source: Panorama of the cassava market - SEBRAE, 2014, p.5).

Looking at the data in the table above, Brazil has the highest household consumption of cassava derivatives such as cassava flour, but the region that contributes most to the consumption of cassava derivatives such as flour and starch is the North.

1.12. Cassava plastic

One solution to plastic is to find an alternative product that is similar in every way to plastic, but less polluting. These biopolymers are obtained from renewable sources, and their properties and characteristics are very similar. These biopolymers are 100% recyclable and take an average of 180 days to decompose.

However, this biopolymer is at an impasse when it comes to large-scale production, as there is a need for more research and a large investment to produce this biopolymer. Although most companies develop this biopolymer from sugar cane, a company in the city of São Carlos-SP has used cassava to produce disposable trays and cups.

4. METHODOLOGY

4.1. Extracting starch from cassava

Materials and reagents:

- Filter paper

- Funnel

- Beaker

- Water

- Blender

- Glass rod.

- Cassava

4.1.1. Method of preparation

Cut the cassava into small cubes and grind them in a blender, after which a paste containing starch, water and pomace is formed. This paste must be strained to separate it from the starchy water. The liquid containing water and starch should be left in a beaker to decant. After decanting, the liquid should be filtered to retain the starch that was present in the water. After filtering, the starch retained in the filter should be taken to the oven to dry.

4.1.2. Process flowchart

The flowchart below shows the main unit operations involved in the cassava starch extraction process, as described above.

Flowchart 1-Extraction of cassava starch

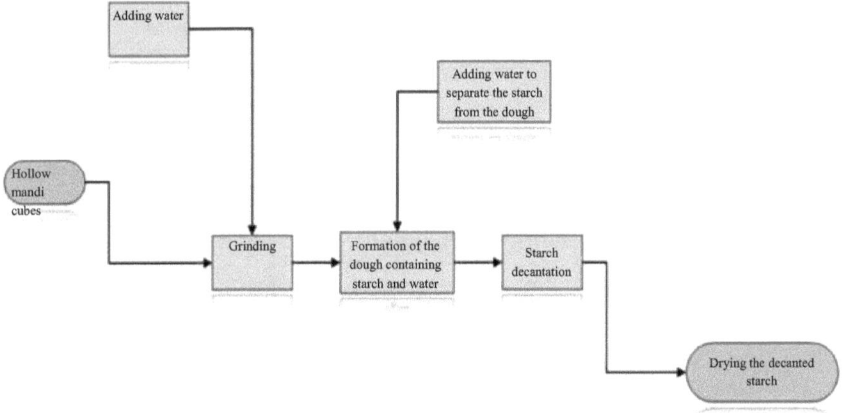

4.2. Biopolymer preparation

Materials and reagents:

- Distilled water

- Previously extracted cassava starch powder

- 0.1M hydrochloric acid

- 0.1M sodium hydroxide

- Glycerin

- Beaker

Glass bastâo

pH tape

4.2.1. Method of preparation

To prepare the biopolymer, 5 grams of cassava starch should be added to a beaker and dissolved in 70 ml of distilled water. After dissolving the starch in the water, this solution should be kept on the heat and stirred for 30 minutes until it comes to the boil. During this heating, 6 ml of 0.1M hydrochloric acid and 4 ml of glycerin should be added. After boiling, the solution should be kept on the heat for another 15 minutes, until the ideal point is reached, thus forming the biopolymer. After the formation of this biopolymer, it should be neutralized with 0.1M sodium hydroxide and transferred to watch glasses for drying.

4.2.2. Process flowchart

Below is a flowchart representing the main unit operations involved in the production process of the biopolymer developed from cassava starch, as described above.

Flowchart 2-Production of biopolymer from cassava starch.

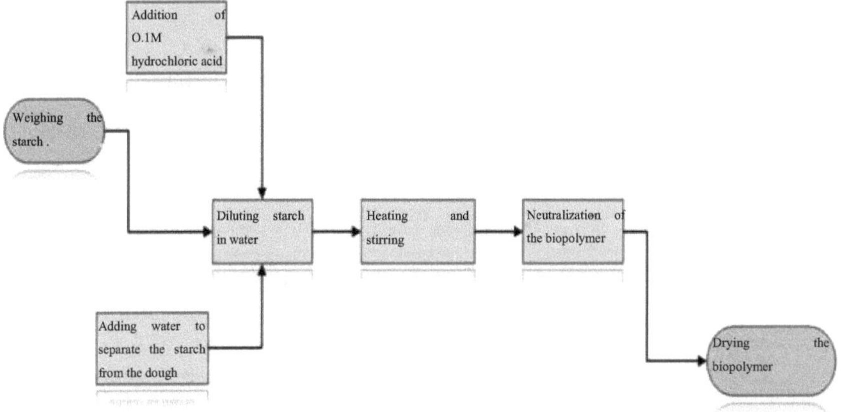

5. Results and discussion

4.3. Starch extraction

During the extraction of the starch, some unit operations were carried out, such as crushing, in which the cassava was crushed together with water in which a thick solution was obtained; this then went through straining where the cassava mass was retained and the solution with the starch was taken to the decanting process, as shown in figure 16 below:

Figure 16 - Starch precipitation

If you look at the picture above, you can see that a white precipitate has formed. This was then filtered. After this unitary operation, the starch was taken to the oven to dry.

4.4. Preparation of the biopolymer

The biopolymer was developed following the methodology below. After the 5 grams of starch were added to a beaker, it was observed that the starch was not solubilized in the water, leaving it with a white colour as shown in figure 20. After being heated and stirred, 6 ml of hydrochloric acid (HCl) and 4 ml of glycerin were added. It was observed that the solution became more turbid in color, which shows that the starch was being broken down by the action of the hydrochloric acid. After 30 minutes of heating, it reached the ideal consistency and was transferred to watch glasses, as shown in figure 21, to be dried.

Figure 17- starch and water

Figure 18 - Starch biopolymer

After drying this biopolymer, decomposition time tests were carried out. Simulations were made of the most common environments in which traditional petroleum plastics are disposed of: water, dry land and wet land.

4.5. Decomposition test

4.5.1. Decomposition in water

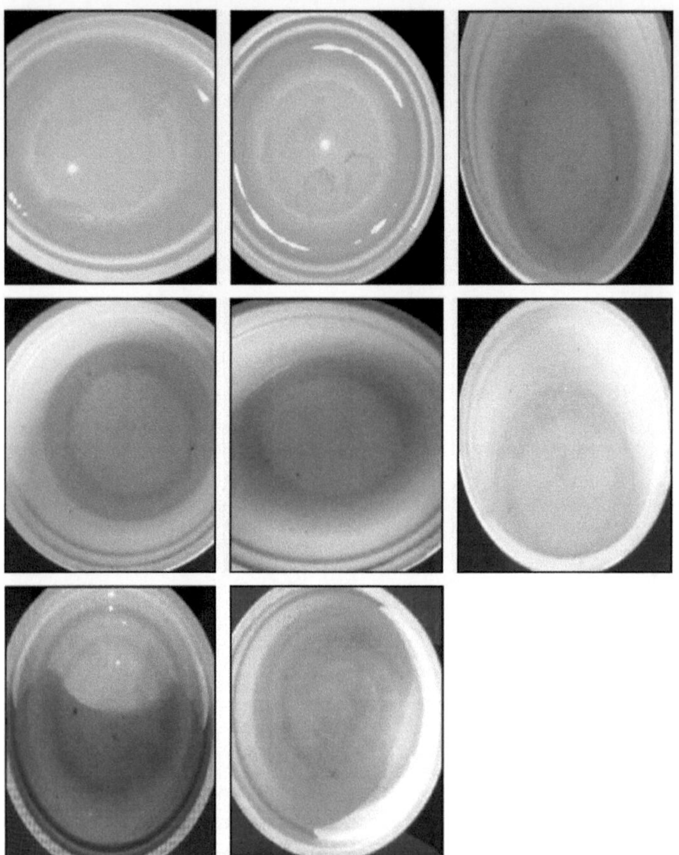

Figure 19-Decomposition of the biopolymer in water

Observing the behavior of the biopolymer in water, in which each figure represents a week of analysis, it can be seen that water acts directly on this biopolymer, since from the second week onwards this polymer began to degrade in the environment, in addition to having undergone changes in its texture, which went from a polymer with a smooth texture to a gelatinous texture.

Below is a comparison of the action of water on the biopolymer:

23/06   11/08

Figure 20-Comparison between the first week and the eighth week of decomposition of the biopolymer in water

In this comparative test, in which eight weeks have passed, the great change in the biopolymer, which was in two medium-sized pieces, can be seen. After six weeks, the biopolymer partially decomposed, leaving only a gelatinous precipitate at the bottom of the container.

4.5.2. Decomposition on dry land

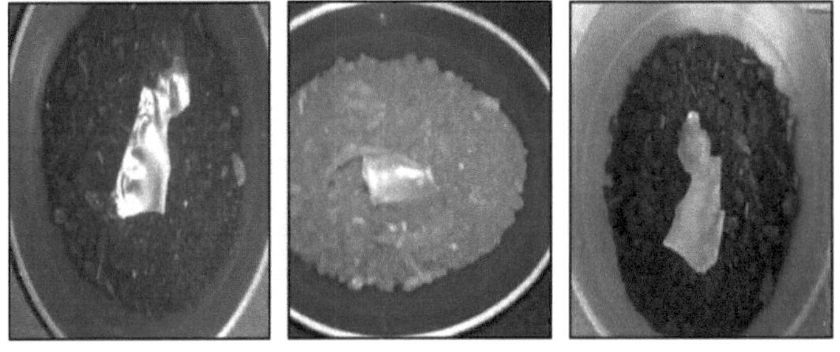

Figure 21-Decomposition of biopolymer on dry land

The biopolymer underwent few changes during its decomposition in this environment. The only changes it underwent were drying out due to the loss of water and it became a biopolymer with a rough, brittle texture.

Below is a comparison of the decomposition process of the biopolymer over a period of eight weeks:

23/0611/08

Figure 22-Comparison between the first week and the eighth week of

decomposition of the biopolymer in dry soil.

If you look at the pictures above, you can clearly see the change in the biopolymer during this period of decomposition. At the start of the test it was flexible and smooth, but after six weeks it became rough and brittle.

4.5.3. Wet decomposition

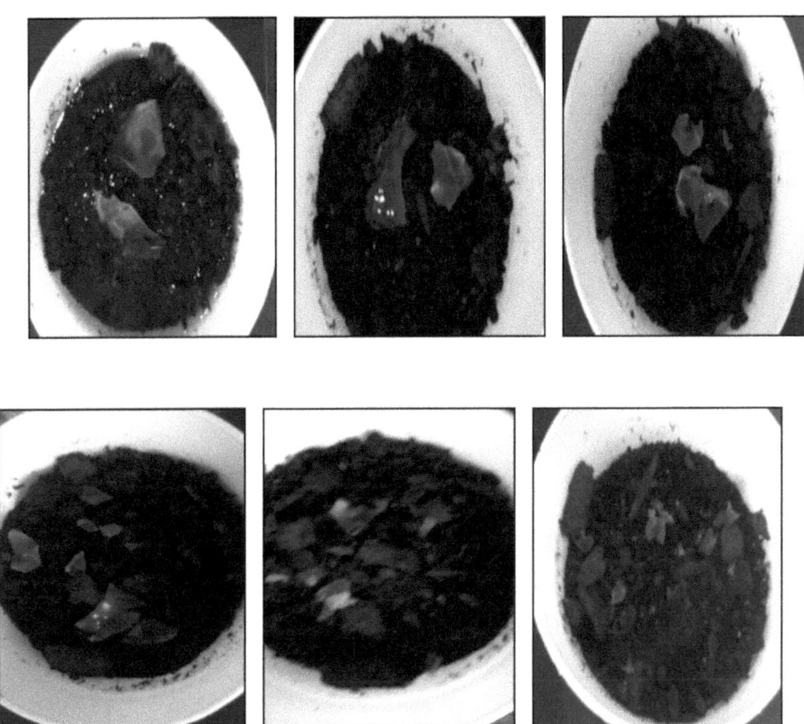

Figure 23-Decomposition of biopolymer in wet soil

This environment was the one that showed the best results, because during the six weeks of testing it was observed that the water and the environment really do influence its decomposition.

Below is a comparison of the decomposition of the biopolymer in the first week and the last week.

23/06   11/08

Figure 24-Comparison between the first week and the eighth week of decomposition of the biopolymer in wet soil

Looking at the figures, it can be seen that this biopolymer decomposed completely in eight weeks, as was seen in the behavior of this biopolymer in water, which also decomposed more quickly. These results coincide with the theoretical data that the biopolymer takes around six weeks to decompose completely.

6. FINAL CONSIDERATIONS

The development of a biopolymer based on starch is a process that is still little known, but which has a promising future, as it is a polymer that decomposes quickly and is produced from a renewable source.

First, the starch extraction process consisted of grinding the cassava and straining the resulting mass. By washing, the water removed the starch, which are small particles, thus promoting greater removal of the starch.

This starch was then decanted and heated for cooking and hydrolysis. During the hydrolysis process, it can be concluded that the starch had indeed broken down, because as soon as the HCl was added, the color of the starch solution changed, thus proving the acid hydrolysis of the starch. During the development of the biopolymer, several factors were observed and some conclusions were drawn from these, such as: as soon as glycerin was added to the hydrolyzed solution mentioned above, this solution changed from a liquid and not very viscous solution to a liquid and very viscous solution, proving that glycerin has gelatinization properties over starch.

After the biopolymer was dried, a decomposition test was carried out, in which the biopolymer was tested in places that simulated the main terrestrial environments. The sites in which it showed the shortest decomposition time were the aquatic site, which simulated disposal in bodies of water, and the wet terrestrial site, which simulated most of the terrestrial environments found. These results show that water has a direct effect on the decomposition of the biopolymer, which took between 6 and 8 weeks to break down.

After analyzing all the results obtained, it can be concluded that the main idea behind the production of the biopolymer, that it would decompose quickly, was achieved, proving that not only is this biopolymer easy to obtain, it is a polymer developed from a renewable source and has high degradability, making it possible to apply it to food packaging, bags and disposable cups.

For those wishing to continue the work, statistical surveys could be carried out to identify the best quantities of HCl and glycerine to add to obtain the best biopolymer, showing them graphically, as well as research into the best types of reactors to use in this process of obtaining the biopolymer, demonstrating the specific conversions of the reaction.

7. REFERENCES

BARROS, C. **Polymer handout. Construction and building materials:** Federal Institute of Education, Science and Technology, South Rio Grande do Sul, Pelotas Campus, 2011.

Newsletter. **Overview of the cassava market**. SEBRAE, 2014.

BONA, J. C. **Preparation and characterization of biodegradable films from blends of starch and polyethylene:** Postgraduate Program in Food Engineering. UFSC - Federal University of Santa Catarina, 2007

BRITO, G.F, et al. **Biopolymers, biodegradable polymers and green polymers:** Electronic journal of materials and processes.

Department of Materials Engineering - Federal University of Campina Grande, 2011.

Brazilian company creates plastic made from cassava, Available at <http://noticias.terra.com.br/ciencia/empresa-brasileira-cria-plastico-feito-com-mandioca,4a47433116492410VgnCLD2000000ec6eb0aRCRD.html > accessed on 25/04/2015

FOGAÇA, J. **Damage caused by oil spills in the oceans,** Available at < http://www.mundoeducacao.com/quimica/danos- causados-por-vazamentos-petroleo-nos-oceanos.htm> accessed on 17/04/2015.

JOHN, L. **How plastic pollutes the oceans and kills marine animals.** Available at < http://viajeaqui.abril.com.br/materias/ilha-de- plastico> accessed on 16/04/2015.

JOHN, L. **How to remove plastic from cassava.** Available at < http://planetasustentavel.abril.com.br/blog/biodiversa/como-tirar-plastico-cassava-272400/> accessed 24/08/2015

MIRANDA, V. A. R. **Blends of polyethylene and modified thermoplastic starch:** Postgraduate Program in Materials Engineering. UFSCAR - Federal

University of São Carlos, 2011.

MORASSI, O. J. **Thermoplastic polymers, thermosets and elastomers.**
Regional Council of Chemistry IV, SP Region. Minicourse 2013.

OLIANI & CERRI. **Pros and cons of plastics for the environment.**
Available at <http://www.ecycle.com.br/component/content/article/35/686-pros-e-contras- do-plastico-para-o-meio-ambiente.html> accessed on 16/04/2015.

OLIANI, S. **Oxybiodegradables: difficult to recycle and alternative to starch.** Available at <
http://www.ecycle.com.br/component/content/article/37-tecnologia-a-favor/726-oxy-degradables-difficult-to-recycle-and-starch-alternative.html >
accessed on 25/04/2015

PASSATORE, C.R. Quimica **dos Polimeros**: 3° Modulo Tècnico em Quimica. Tiquatira State Technical School - Paula Souza Center, 2013.

PIATTI.T.M & RODRIGUES. R. A. F. **Plastics: Characteristics, uses, production and environmental impacts.** UFAL - Federal University of Alagoas, 2005.

PLA: The compostable plastic, Available at<
http://www.ecycle.com.br/component/content/article/37/738-pla-o-plastico-compostavel.html

Almost 270,000 tons of plastic pollute the oceans, says study, VEJA. Available at
<http://veja.abril.com.br/noticia/ciencia/quase-270-mil-toneladas-de-plastico-poluem-os-oceanos-diz-estudo/ > access 21/08/2015

SPIER, F. **Effect of alkaline, acid and oxidative treatments on the properties of corn starch:** Postgraduate program in agro-industrial science and technology. UFPEL - Federal University of Pelotas, 2010.

Yamaoka, M. **The cassava that turns into** cups<

http://planetasustentavel.abril.com.br/noticia/lixo/empresa-investir-evitar-
poluicao-mandioca-vira-copinhos-682300.shtml> access 01/10/2015

yes

I want morebooks!

Buy your books fast and straightforward online - at one of world's fastest growing online book stores! Environmentally sound due to Print-on-Demand technologies.

Buy your books online at
www.morebooks.shop

Kaufen Sie Ihre Bücher schnell und unkompliziert online – auf einer der am schnellsten wachsenden Buchhandelsplattformen weltweit! Dank Print-On-Demand umwelt- und ressourcenschonend produziert.

Bücher schneller online kaufen
www.morebooks.shop